Number and Operations

Double the Number

Dolphins live in the ocean.
They have fins.
Dolphins swim fast.

How many fins on each **side**?

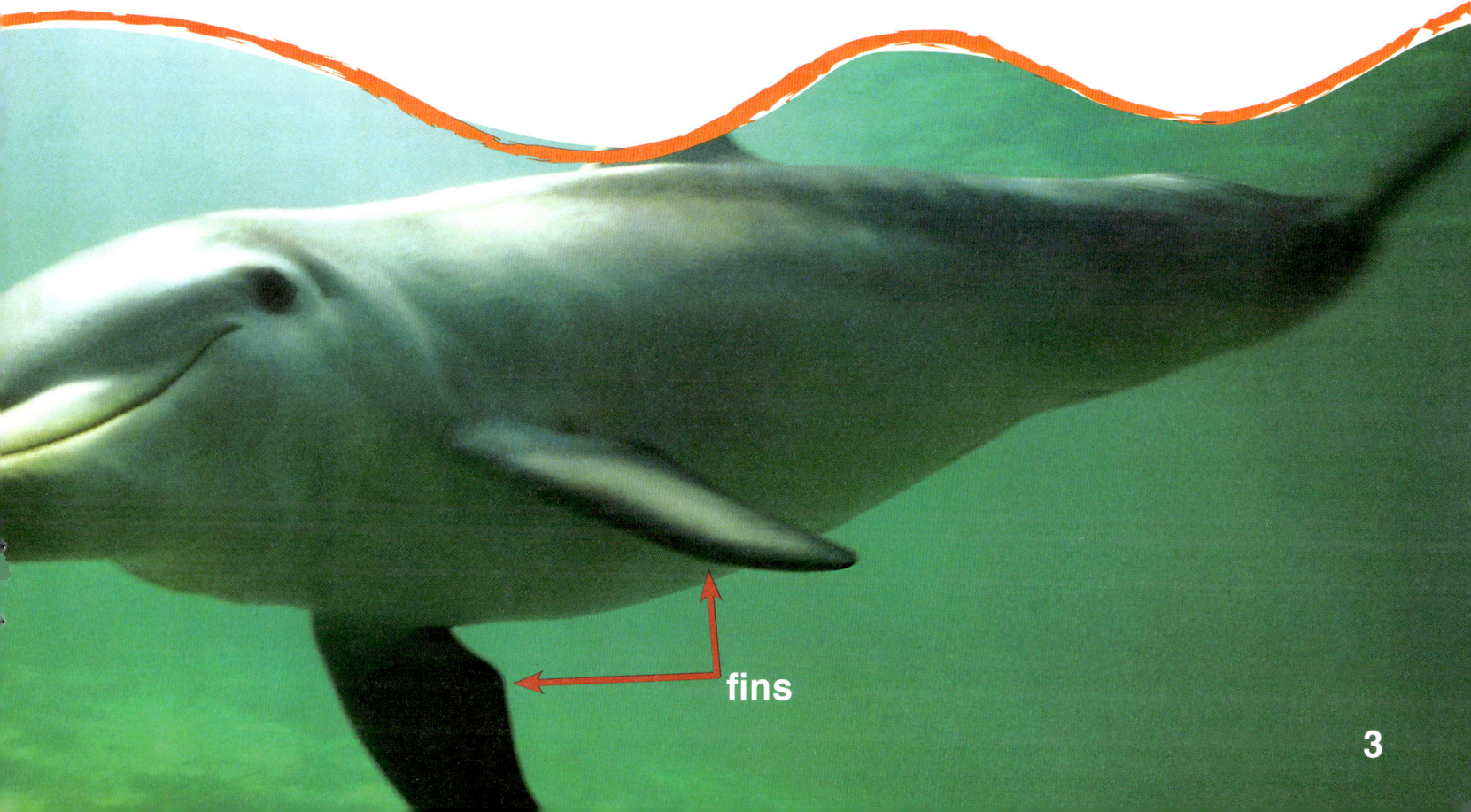

fins

Spiders live in many places.
Some make silk.
Most spiders spin webs.

Count the legs on one side.
How many legs **in all**?

Did You Know?

Some spiders have 12 eyes!

Polar bears live in cold places.
They have big paws.
Paws help polar bears stay warm.

Count the pads on one paw.
How many pads on two paws?

pads

Geckos are good diggers.
Count the toes on each foot.
How many toes **in all**?